Printed in Mexico

ISBN 978-0-15-362192-5
ISBN 0-15-362192-3

2 3 4 5 6 7 8 9 10 050 16 15 14 13 12 11 10 09 08

Visit *The Learning Site!*
**www.harcourtschool.com**

# Heat Energy

**Heat** is energy that makes things hot. The sun is what gives Earth heat energy.

The sun is a star. It warms the land, air, and water all around you.

**Fast Fact**

**The middle of the sun is at least 15 million degrees Celsius (27 million degrees Fahrenheit).**

**The sun gives off heat energy to warm the sand, the water, and the air.**

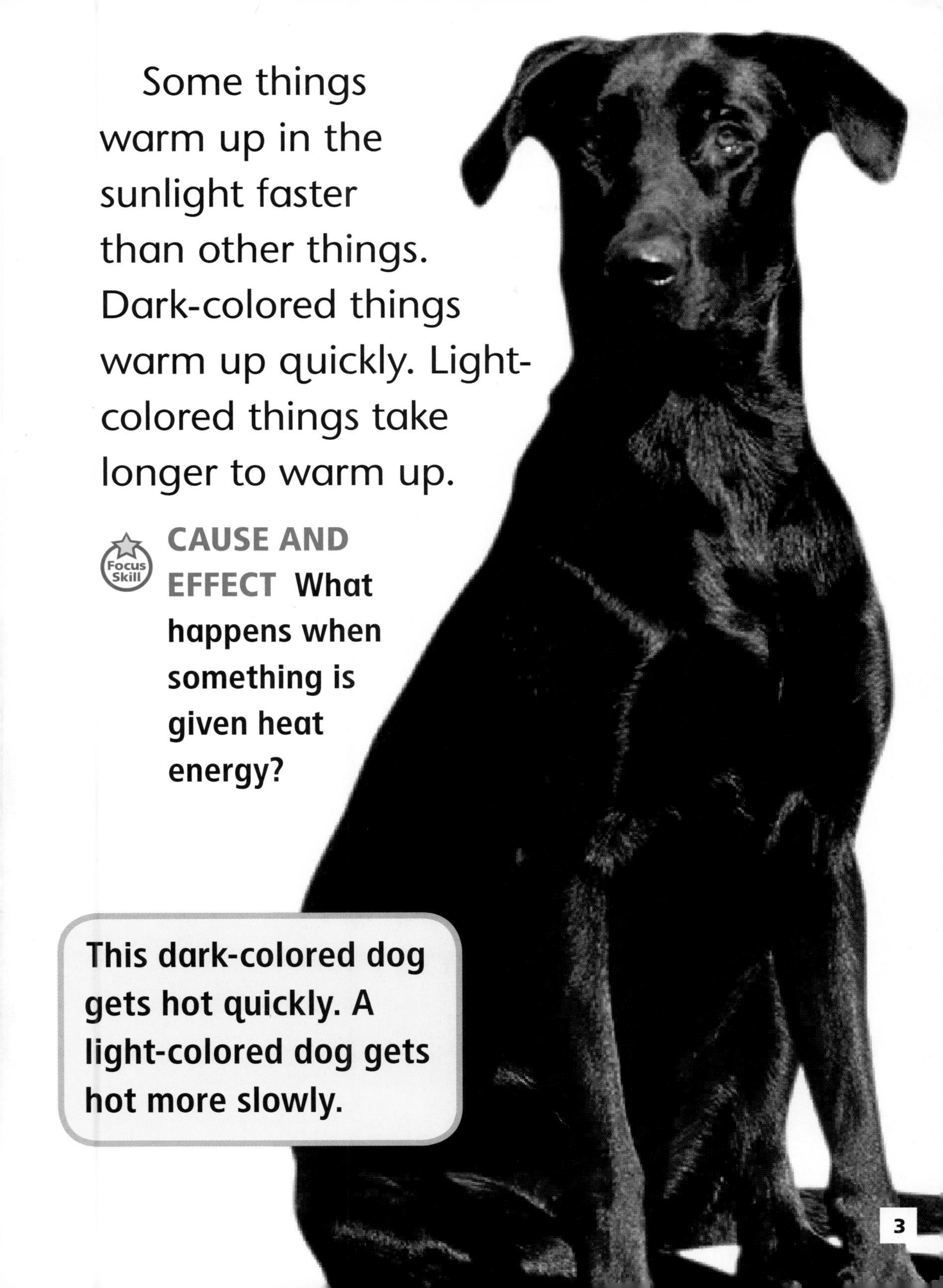

Some things warm up in the sunlight faster than other things. Dark-colored things warm up quickly. Light-colored things take longer to warm up.

Focus Skill **CAUSE AND EFFECT** **What happens when something is given heat energy?**

**This dark-colored dog gets hot quickly. A light-colored dog gets hot more slowly.**

## Feeling the Heat

You feel heat from the sun. You can feel heat from other things, too.

Fire makes things warm. You can feel heat when something is burned.

Fire gives off heat. It warms this room.

Sometimes light bends. The bending of light as it moves from one material to another is called **refraction**.

If you put part of your finger in water, you can see the difference between reflection and refraction. Your finger above the water looks the same as you always see it. Light bounces off it straight to your eyes.

But the part of your finger below the water might seem bent or even disconnected from the rest of your finger! That's because light bends when it goes through the water.

You see the part of your finger above the water by reflection. You see the part of your finger below the water by refraction. The bending light makes it look like your finger is not connected below the water.

**When light moves from air to the water, it bends. The straw looks bent below the surface of the water.**

**SEQUENCE What happens to light after it hits water?**

## Light Energy

**Light** is a kind of energy. The sun gives light energy for the world around us.

When you go outside, you can see trees and houses. Light helps you to see.

The sun lights up the world around us.

A candle is burned to give light.

It can get dark at night. It can be hard to see. Fires and lamps give off light. People use fires and lamps to help them see.

**MAIN IDEA AND DETAILS**
**Where can we get light?**

**Fast Fact**

**A 100 watt light bulb at your house makes as much light as about 1,500 candles!**

# Light on the Move

Light moves. It passes through clear objects. It passes through clear glass. It passes through clear air.

**These windows are made of clear glass. Light passes through them.**

Objects that are not clear, block light. Trees, buildings, and even people can block light. They can make shadows. A **shadow** is a dark place made when an object blocks light. You can see lots of shadows on a sunny day.

**CAUSE AND EFFECT What causes a tree to make a shadow?**

**Can you see shadows? What blocked the light from passing through?**

# Sound Energy

Did you know that sound is a kind of energy? **Sound** is energy you can hear. Music, talking, and a dog barking are all sound. Sound is all around us.

**Musical instruments make many different sounds.**

**A fire engine gives light energy from its lights. It gives sound energy from its siren.**

**COMPARE AND CONTRAST** **How is the light from the sun like sound from a drum?**

## Summary

The sun sends heat and light energy to Earth. We also get heat and light in other ways. Light can pass through some things, but not others. Sound is a kind of energy, too.

# Glossary

**heat** A kind of energy that makes things hot (2, 3, 4, 5, 11)

**light** A kind of energy that helps us see (6, 7, 8, 9, 11)

**shadow** A dark place made when an object blocks light (9)

**sound** A kind of energy that you can hear (10, 11)